— *of* —

MARYLAND, DELAWARE

&

VIRGINIA

(including Washington, DC)

Tamara Eder

with contributions from Ian Sheldon

LONE PINE

© 2001 by Lone Pine Publishing
First printed in 2001 10 9 8 7 6 5 4 3 2 1
Printed in Canada

THE PUBLISHER: LONE PINE PUBLISHING

1901 Raymond Avenue SW, Suite C	10145-81 Avenue
Renton, WA 98055	Edmonton, AB T6E 1W9
USA	Canada

Website: http://www.lonepinepublishing.com

National Library of Canada Cataloguing in Publication Data

Eder, Tamara, (date)
  Animal tracks of Maryland, Delaware & Virginia

  Includes bibliographical references and index.
  ISBN 1-55105-309-8

  1. Animal tracks—Virginia—Identification. 2. Animal tracks—Middle Atlantic
States—Identification. I. Title.
QL768.E358 2001   591.47'9   C2001-910535-5

Editorial Director: Nancy Foulds
Editor: Volker Bodegom
Proofreader: Lee Craig
Production Manager: Jody Reekie
Design, Layout & Production: Volker Bodegom, Monica Triska
Cover Design: Robert Weidemann
Technical Contributor: Mark Elbroch
Cartography: Volker Bodegom
Animal Illustrations: by Gary Ross, except for those by Kindrie Grove
(pp. 9, 91), Ewa Pluciennik (p. 111) and Ian Sheldon (p. 123).
Track Illustrations: Ian Sheldon
Cover Illustration: Raccoon by Gary Ross
Scanning: Elite Lithographers Ltd.

We acknowledge the financial support of the Government of Canada
through the Book Publishing Industry Development Program (BPIDP)
for our publishing activities.

PC: P4

CONTENTS

INTRODUCTION

If you have ever spent time with an experienced tracker, or perhaps a veteran hunter, then you know just how much there is to learn about the subject of tracking and just how exciting the challenge of tracking animals can be. Maybe you think that tracking is no fun, because all you get to see is the animal's prints. What about the animal itself—is that not much more exciting? Well, for most of us who don't spend a great deal of time in the beautiful wilderness of Virginia, Maryland and Delaware, the chances of seeing the elusive Mink or the cunning Red Fox are slim. The closest that we may ever get to some animals will be through their tracks, and they can inspire a very intimate experience. Remember, you are following in the footsteps of the unseen—animals that are in pursuit of prey, or perhaps being pursued as prey.

This book offers an introduction to the complex world of tracking animals. Sometimes tracking is easy. At other times it is an incredible challenge that leaves you wondering just what animal made those unusual tracks. Take this book into the field with you, and it can provide some help with the first steps to identification. Animal tracks and trails are this book's focus; you will learn to recognize subtle differences for both. There are, of course, many additional signs to consider, such as scat and food caches, all of which help you to understand the animal that you are tracking.

It takes many years to become an expert tracker. Tracking is one of those skills that grows with you as you acquire new knowledge in new situations. Most importantly, you will have an intimate experience with nature. You will learn the secrets of the seldom seen. The more you discover, the more you will want to know. And, by developing a good understanding of tracking, you will gain an excellent appreciation of the intricacies and delights of the marvelous natural world.

How to Use This Book

Most importantly, take this book into the field with you! Relying on your memory is not an adequate way to identify tracks. Track identification has to be done in the field, or with detailed sketches and notes that you can take home. Much of the process of identification involves circumstantial evidence, so you will have much more success when standing beside the track.

This book is laid out in an easy-to-use format. Beginning on p. 126, there is a quick reference appendix to the tracks of all the animals illustrated in the book. This appendix is a fast way to familiarize yourself with certain tracks, and it guides you to the more informative descriptions of each animal.

Each animal's description is illustrated with the appropriate footprints and the track patterns that it usually leaves.

Mink

Although these illustrations are not exhaustive, they do show the tracks or groups of prints that you will most likely see. You will find a list of dimensions for the tracks, giving the general range, but there will always be extremes, just as there are with people who have unusually small or large feet. Under the category 'Size' (of animal), the 'greater-than' sign (>) is used when the size difference between the sexes is pronounced.

If you think that you may have identified a track, check the 'Similar Species' section. This section is designed to help you confirm your conclusions by pointing out other animals that leave similar tracks and by showing you ways to distinguish among them.

As you read this book, you will notice an abundance of words such as 'often,' 'mostly' and 'usually.' Unfortunately, tracking will never be an exact science; we cannot expect animals to conform to our expectations, so be prepared for the unpredictable.

Tips on Tracking

As you flip through this guide, you will notice clear, well-formed prints. Do not be deceived! It is a rare track that will ever show so clearly. For a good, clear print, the perfect conditions are slightly wet, shallow snow that isn't melting, or slightly soft mud that isn't actually wet. These conditions can be rare—most often you will be dealing with incomplete or faint prints, where you cannot even really be sure of the number of toes.

Should you find yourself looking at a clear print, then the job of identification is much easier. There are

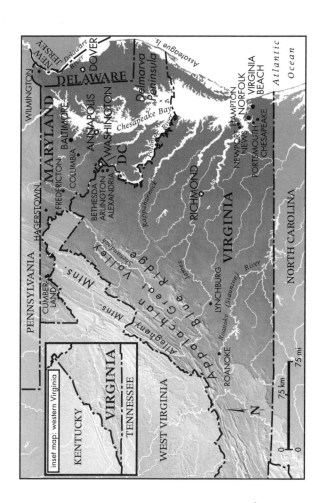

7

a number of key features to look for: measure the length and width of the print, count the number of toes, check for claw marks and note how far away they are from the body of the print, and look for a heel mark. Keep in mind more subtle features, such as the spacing between the toes, whether or not they are parallel, and whether fur on the sole of the foot has made the print less clear.

When you are faced with the challenge of identifying an unclear print—or even if you think that you have made a successful identification from one print alone—look beyond the single footprint and search out others. Do not rely on the dimensions of just one print, but collect measurements from several prints to get an average impression. Even the prints within one trail can show a lot of variation.

Try to determine which is the fore print and which is the hind, and remember that many animals are built very differently from humans, having larger forefeet than hind feet. Sometimes the prints will overlap, or they can be directly on top of one another in a direct register. For some animals, the fore and hind prints are pretty much the same.

Check out the pattern that the tracks make together in the trail and follow the trail for as many paces as is necessary for you to become familiar with the pattern.

Patterns are very important and can be the distinguishing feature between different animals with otherwise similar tracks.

Follow the trail for some distance—it can give you some vital clues. For example, the trail may lead you to a tree, indicating that the animal is a climber, or it may lead down into a burrow. This part of tracking can be the most rewarding, because you are following the life of the animal as it hunts, runs, walks, jumps, feeds or tries to escape a predator.

Take into consideration the habitat. Sometimes habitat alone will allow you to distinguish very similar tracks—one species might be found on riverbanks, whereas another might be encountered just in dense forest.

Think about your geographical location, too, because some animals have a limited range. This consideration can rule out some species and help you with your identification.

Remember that every animal will at some point leave a print or trail that looks just like the print or trail of a completely different animal!

Finally, keep in mind that if you track quietly, you might catch up with the maker of the prints.

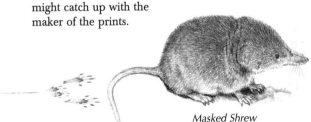

Masked Shrew

Terms & Measurements

Some of the terms used in tracking can be rather confusing, and they often depend on personal interpretation. For example, what comes to your mind if you see the word 'hopping'? Perhaps you see a person hopping about on one leg, or perhaps you see a rabbit hopping through the countryside. Clearly, one person's perception of motion can be very different from another's. Some useful terms are explained on the next few pages, to clarify what is meant in this book, and, where appropriate, how the measurements given fit in with each term.

The following terms are sometimes used loosely and interchangeably—for example, a rabbit might be described as 'a hopper' and a squirrel as 'a bounder,' yet both leave the same pattern of prints in the same sequence.

Ambling: Fast, rolling walking.

hind — 🐾 🐾 — fore
print   print

Bounding: A gait of four-legged animals in which the two hind feet land simultaneously, usually registering in front of the fore prints. It is common in rodents and the rabbit family. 'Hopping' or 'jumping' can often be substituted.

Gait: Describes how an animal is moving at some point in time. Different gaits result in different observable trail characteristics.

Galloping: A gait used by animals with four legs of even length, such as dogs, moving at high speed, hind feet registering in front of forefeet.

hind prints   *fore prints*   *gallop group*

Hopping: Similar to bounding. With four-legged animals, it is usually indicated by tight clusters of prints, fore prints set between and behind the hind prints. A bird hopping on two feet creates a series of paired tracks along its trail.

Loping: Like galloping but slower, with each foot falling independently and leaving a trail pattern that consists of groups of tracks in the sequence fore-hind-fore-hind, usually roughly in a line.

11

Mustelids (weasel family) often use **2×2 loping**, in which the hind feet register directly on the fore prints. The resulting pattern has angled, paired tracks.

Running: Like galloping, but applied generally to animals moving at high speed. Also used for two-legged animals.

Trotting: Faster than walking, slower than running. The diagonally opposite limbs move simultaneously; that is, the right forefoot with the left hind, then the left forefoot with the right hind. This gait is the natural one for canids (dog family), short-tailed shrews and voles.

Canids may use **side-trotting**, a fast trotting in which the hind end of the animal shifts to one side. The resulting track pattern has paired tracks, with all the fore prints on one side and all the hind prints on the other.

hind print →

fore print →

Walking: A slow gait in which each foot moves independently of the others, resulting in an alternating track pattern. This gait is common for felines (cat family) and deer, as well as wide-bodied animals, such as bears and porcupines. The term is also used for two-legged animals.

Other Tracking Terms:

Dewclaws: Two small, toe-like structures set above and behind the main foot of most hoofed animals.

Direct Register: The hind foot falls directly on the fore print.

double register   direct register

Double Register: The hind foot registers, overlapping the fore print only slightly or falling beside it, so that both prints can be seen at least in part.

Dragline: A line left in snow or mud by a foot or the tail dragging over the surface.

dragline

Gallop Group: A track pattern of four prints made at a gallop, usually with the hind feet registering in front of the forefeet (see '***galloping***' for illustration).

Height: Taken at the animal's shoulder.

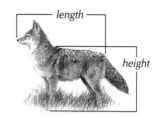

Length: The animal's body length from head to rump, not including the tail, unless otherwise indicated.

Metacarpal Pad: A small pad near the palm pad or between the palm pad and heel on the forefeet of bears and members of the weasel family.

Print (also called '**track**'): Fore and hind prints are treated individually. Print dimensions given are 'length' (including claws—maximum values may represent occasional heel register for some animals) and 'width.' A group of prints made by each of the animal's feet makes up a track pattern.

Register: To leave a mark—said about a foot, claw or other part of an animal's body.

Retractable: Describes claws that can be pulled in to keep them sharp, as with the cat family; these claws do not register in the prints. Foxes have semi-retractable claws.

Sitzmark: The mark left on the ground by an animal falling or jumping from a tree.

Straddle: The total width of the trail, all prints considered.

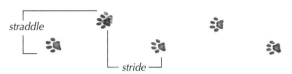

straddle

stride

Stride: For consistency among different animals, the stride is taken as the distance from the center of one print (or print group) to the center of the next one. Some books may use the term 'pace.'

Track: Same as '***print***.'

Track Pattern: The pattern left after each foot registers once; a set of prints, such as a gallop group.

Trail: A series of track patterns; think of it as the path of the animal.

Red Fox

MAMMALS

River Otter

White-tailed Deer

Fore and Hind Prints
Length: 2–3.5 in (5–9 cm)
Width: 1.6–2.5 in (4–6.5 cm)

Straddle
5–10 in (13–25 cm)

Stride
Walking: 10–20 in (25–50 cm)
Galloping: 6–15 ft (1.8–4.5 m)

Size (buck>doe)
Height: 3–3.5 ft (90–110 cm)
Length: to 6.3 ft (1.9 m)

Weight
120–350 lb (55–160 kg)

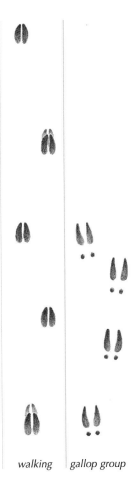

walking   *gallop group*

WHITE-TAILED DEER
Odocoileus virginianus

The keen hearing of this deer guarantees that it knows about you before you know about it. Frequently, all that we see is its conspicuous white tail in the distance as it gallops away, earning this deer the nickname 'Flagtail.' This adaptable deer may be found throughout the state in small groups at the edges of forests and in brushlands. The White-tailed Deer can be common around ranches and residential areas.

This deer's prints are heart-shaped and pointed. Its alternating walking track pattern shows the hind prints direct registered or double registered on the fore prints. When a deer gallops on a soft surface, such as snow, the dewclaws register. This flighty deer gallops in the usual style, leaving hind prints ahead of fore prints, with the toes splayed wide for steadier, safer footing.

Similar Species: The Sika Deer (*Cervus nippon*) makes similar but smaller prints. In farming areas the prints of domesticated Elk (*C. elaphus*) or the Domestic Pig (*Sus scrofa*) could be mistaken for deer prints.

Horse

Fore Print
(hind print is slightly smaller)
Length: 4.5–6 in (11–15 cm)
Width: 4.5–5.5 in (11–14 cm)
Straddle
2–7.5 in (5–19 cm)
Stride
Walking: 17–28 in (43–70 cm)
Size
Height: to 6 ft (1.8 m)
Weight
to 1500 lb (680 kg)

walking

HORSE
Equus caballus

Outdoor adventures on horseback are a popular activity, so you can expect horse tracks to show up almost anywhere. In some places, such as Assateague Island, feral Horses (Assateague Ponies) are abundant.

Unlike any other animal in this book, the Horse has just one huge toe on each foot. This toe leaves an oval print with a distinctive 'frog' (V-shaped mark) at its base. If a Horse is shod, the horseshoe shows up clearly as a firm wall at the outside of the print. Not all horses are shod, so do not expect to see this outer wall on every horse print. A typical, unhurried horse trail shows an alternating walking pattern, with the hind prints registered on or behind the slightly larger fore prints. Horses are capable of a range of speeds—up to a full gallop—but most recreational horseback riders take a more leisurely outlook on life, preferring to walk their horses and soak up the beautiful scenery!

Similar Species: Mules (rarely shod) have smaller tracks.

Black Bear

fore

hind

Fore Print
Length: 4–6.3 in (10–16 cm)
Width: 3.8–5.5 in (9.5–14 cm)

Hind Print
Length: 6–7 in (15–18 cm)
Width: 3.5–5.5 in (9–14 cm)

Straddle
9–15 in (23–38 cm)

Stride
Walking: 17–23 in (43–58 cm)

Size (male>female)
Height: 3–3.5 ft (90–110 cm)
Length: 5–6 ft (1.5–1.8 m)

Weight
200–600 lb (90–270 kg)

walking

BLACK BEAR
Ursus americanus

The Black Bear has a scattered range that includes the forested mountain regions of western Maryland and western Virginia, but do not expect to see its tracks in winter when it hibernates. Finding fresh bear tracks can be a thrill, but take care—the bear may be just ahead. Never underestimate the potential power of a surprised bear!

Bear prints somewhat resemble small human prints, but they are wider and show claw marks. The small inner toe rarely registers. The forefoot's small heel pad often registers, and the hind print shows a big heel. The bear's slow walk results in a slightly pigeon-toed double register, with the hind print on the fore print. More frequently, at a faster pace, the hind foot oversteps the forefoot. When a bear runs, the two hind feet register in front of the forefeet in an extended cluster. Along well-worn bear paths, look for 'digs' (patches of dug-up earth) and 'bear trees' whose scratched bark shows that this bear climbs.

Similar Species: No other wild animal is likely to leave similar tracks in this area.

Domestic Dog

fore

hind

Fore Print
(hind print is smaller)
Length: 1–5.5 in (2.5–14 cm)
Width: 1–5 in (2.5–13 cm)

Straddle
1.5–8 in (3.8–20 cm)

Stride
Walking: 3–32 in (7.5–80 cm)
Loping to Galloping: to 9 ft (2.7 m)

Size
Very variable

Weight
Very variable

walking

loping to galloping

DOMESTIC DOG
Canis familiaris

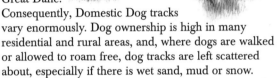

Dogs come in many shapes and sizes, from the tiny Chihuahua with its dainty feet to the robust and powerful Great Dane. Consequently, Domestic Dog tracks vary enormously. Dog ownership is high in many residential and rural areas, and, where dogs are walked or allowed to roam free, dog tracks are left scattered about, especially if there is wet sand, mud or snow.

The forefeet of the Domestic Dog, which are much larger that the hind feet and support more of the animal's weight, leave the clearest tracks. When a dog walks, the hind prints usually register ahead of or beside the fore prints. As the dog moves faster, it trots and then lopes before it gallops. In a trot or lope pattern the prints alternate fore-hind-fore-hind, whereas a gallop group shows (from back to front) fore-fore-hind-hind.

Similar Species: Keep in mind that dog prints are usually found close to human tracks or activity. Fox (pp. 28–31) prints may be confused with small dog prints. Coyote (p. 26) prints, usually more oval and splaying less, will be in a more direct trail.

Coyote

fore

hind

Fore Print
(hind print is slightly smaller)
Length: 2.4–3.2 in (6–8 cm)
Width: 1.6–2.4 in (4–6 cm)

Straddle
4–7 in (10–18 cm)

Stride
Walking: 8–16 in (20–40 cm)
Trotting: 17–23 (43–58 cm)
Galloping/Leaping:
    2.5–10 ft (0.8–3 m)

Size (female is slightly smaller)
Height: 23–26 in (58–65 cm)
Length: 32–40 in (80–100 cm)

Weight
20–50 lb (9–23 kg)

walking
or trotting

gallop
group

COYOTE
(Brush Wolf, Prairie Wolf)
Canis latrans

This widespread, adaptable canine prefers open grasslands or woodlands. On its own, with a mate or in a family pack, it hunts rodents and larger prey. If you find a coyote den—usually a wide-mouthed tunnel leading into a nesting chamber—do not bother the family or the female will have to move her pups to a safer location.

The hind print is slightly smaller than the oval fore print, and its less triangular heel pad rarely registers clearly. The claws of the two outer toes usually do not register. The Coyote typically walks or trots in an alternating pattern; the walk has a wider straddle, and the trotting trail is often very straight. When the Coyote gallops, its hind feet fall in front of its forefeet; the faster it goes, the straighter the gallop group. The Coyote's tail, which hangs down, leaves a dragline in deep snow.

Similar Species: A Domestic Dog's (p. 24) less oval prints splay more, and its trail will be erratic. Foot hairs blur Red Fox (p. 28) prints (usually smaller). Gray Fox (p. 30) prints are much smaller.

Red Fox

fore

hind

Fore Print
(hind print is slightly smaller)
Length: 2.1–3 in (5.3–7.5 cm)
Width: 1.6–2.3 in (4–5.8 cm)

Straddle
2–3.5 in (5–9 cm)

Stride
Trotting: 12–18 in (30–45 cm)
Side-trotting: 14–21 in (35–53 cm)

Size (vixen is slightly smaller)
Height: 14 in (35 cm)
Length: 22–25 in (55–65 cm)

Weight
7–15 lb (3.2–7 kg)

trotting *side-trotting*

RED FOX
Vulpes vulpes

Very adaptable and intelligent, this beautiful and notoriously cunning fox is found throughout the region in a variety of habitats ranging from forests to open areas.

Abundant foot hair permits only parts of the toes and heel pads to register, but with no fine detail. The horizontal or slightly curved bar across the fore heel pad is diagnostic. A trotting Red Fox leaves a distinctive straight alternating trail—the hind print direct registers on the wider fore print. When the fox side-trots, its print pairs show the hind print to one side of the fore print in typical canid fashion. This fox gallops like the Coyote (p. 26). The faster the gallop, the straighter the gallop group.

Similar Species: Other canid prints lack the bar across the fore heel pad. Domestic Dog (p. 24) prints can be of similar size, but they will have a shorter stride and a less direct trail. Small Coyote prints are similar, but they will have a wider straddle, and the toe marks will be more bulbous. The Gray Fox (p. 30) makes smaller prints.

Gray Fox

fore

hind

Fore Print
(hind print is slightly smaller)
Length: 1.3–2.1 in (3.3–5.3 cm)
Width: 1.1–1.5 in (2.8–3.8 cm)
Straddle
2–4 in (5–10 cm)
Stride
Walking/Trotting: 7–12 in (18–30 cm)
Size
Height: 14 in (35 cm)
Length: 21–30 in (53–75 cm)
Weight
7–15 lb (3.2–7 kg)

walking

GRAY FOX
Urocyon cinereoargenteus

This small, shy fox is widespread but quite rare; look for its tracks in woodlands and chaparral country. The Gray Fox is the only fox that climbs trees, which it does either for safety or to forage.

The forefoot registers better than the smaller hind foot, and the hind foot's long, semi-retractable claws do not always register. The heel pads are often unclear—they sometimes show up just as small, round dots. When it walks, this fox leaves a neat alternating track pattern. When it trots, its prints fall in pairs, with the fore print set diagonally behind the hind print. The Gray Fox's gallop group is like the Coyote's (p. 26).

Similar Species: The Red Fox's (p. 28) fore heel pad has a bar across it; in general, Red Fox prints are larger and less clear (because of thick fur), and they have a longer stride and narrower straddle. Coyote tracks are much larger. Domestic Dog (p. 24) prints are often larger. Feline (pp. 32–37) prints lack claw marks, and they have larger, less symmetrical heel pads.

Mountain Lion

fore

hind

Fore Print
(hind print is slightly smaller)
Length: 3–4.5 in (7.5–11 cm)
Width: 3.3–4.8 in (8.5–12 cm)
Straddle
8–12 in (20–30 cm)
Stride
Walking: 13–32 in (33–80 cm)
Bounding: to 12 ft (3.7 m)
Size
Height: 25–32 in (65–80 cm)
Length: 3.5–5 ft (1.1–1.5 m)
Weight
70–200 lb (32–90 kg)

walking

MOUNTAIN LION
(Panther, Cougar, Puma)
Puma concolor

Shy, elusive and nocturnal, the Mountain Lion is spread widely but sparsely because of its need for a big home territory. Once common in the eastern part of the country, it is now extirpated from most of its historic range. Nevertheless, it remains on the endangered species list for this region, and infrequent sightings are still reported.

Mountain Lion prints tend to be wider than long. The retractable claws never register. Thick foot fur enlarges the print in winter and may stop the two lobes on the front of the heel pad from registering clearly. In the walking gait, the hind print direct registers or double registers on the larger fore print. As the pace increases, the hind print tends to fall ahead of the fore print. In snow, the thick, long tail may leave a dragline that can blur print detail. A Mountain Lion seldom gallops, but it is capable of long bounds. Also look for partly buried scat and kills covered for later eating.

Similar Species: Large Bobcat (p. 34) prints may be confused with juvenile Mountain Lion prints. Coyote (p. 26) tracks show claw marks.

Bobcat

fore

hind

**Fore Print
(hind print is slightly smaller)**
Length: 1.8–2.5 in (4.5–6.5 cm)
Width: 1.8–2.5 in (4.5–6.5 cm)

Straddle
4–7 in (10–18 cm)

Stride
Walking: 8–16 in (20–40 cm)
Running: 4–8 ft (1.2–2.4 m)

Size (female is slightly smaller)
Height: 20–22 in (50–55 cm)
Length: 25–30 in (65–75 cm)

Weight
15–35 lb (7–16 kg)

walking

*ambling
to loping*

BOBCAT
(Wildcat)
Lynx rufus

The handsome Bobcat, a stealthy and usually nocturnal hunter in western Virginia and western Maryland, is seldom seen. Very adaptable, it can leave tracks anywhere from wild mountainsides to chaparral and even in residential areas.

A walking Bobcat's hind feet usually register directly on its larger fore prints. As the Bobcat picks up speed, its trail becomes an ambling pattern of paired prints, the hind leading the fore. At even greater speeds, it leaves four-print groups in a lope pattern. The fore prints especially show asymmetry. The front part of the heel pad has two lobes and the rear part has three. In deep snow the Bobcat's feet leave draglines. Half-buried scat along the Bobcat's meandering trail marks its territory.

Similar Species: Juvenile Mountain Lion (p. 32) prints may resemble large Bobcat prints. A large Domestic Cat (p. 36) will make similar prints with a shorter stride and a narrower straddle. Fox (pp. 28–31), Domestic Dog (p. 24) and Coyote (p. 26) prints are narrower than they are long and show claw marks; the fronts of their footpads are once-lobed. Some mustelid (pp. 46–49) prints may look alike.

Domestic Cat

fore

hind

Fore Print
(hind print is slightly smaller)
Length: 1–1.6 in (2.5–4 cm)
Width: 1–1.8 in (2.5–4.5 cm)

Straddle
2.4–4.5 in (6–11 cm)

Stride
Walking: 5–8 in (13–20 cm)
Loping/Galloping:
    14–32 in (35–80 cm)

Size (male>female)
Height: 20–22 in (50–55 cm)
Length with tail: 30 in (75 cm)

Weight
6.5–13 lb (3–6 kg)

walking

loping to galloping

DOMESTIC CAT
(House Cat)
Felis catus

   The tracks of the familiar and abundant Domestic Cat can show up almost any place where there are people. Abandoned cats may roam farther afield; these 'feral cats' lead a pretty wild and independent existence. Domestic Cats come in many shapes, sizes and colors.

   As with all felines, a Domestic Cat's fore print and slightly smaller hind print both show four toe pads. Its retractable claws, kept clean and sharp for catching prey, do not register. Cat prints usually show a slight asymmetry, with one toe leading the others. A Domestic Cat makes a neat alternating walking track pattern, usually in direct register, as one would expect from this animal's fastidious nature. When a cat picks up speed, it leaves clusters of four prints, the hind feet registering in front of the forefeet.

Similar Species: A small Bobcat (p. 34) may leave tracks similar to a very large Domestic Cat's. Domestic Dog (p. 24) and Fox (pp. 28–31) prints show claw marks. Weasel (pp. 48–51) or small Mink (p. 46) prints that show four toes may also seem similar; look for mustelid habits.

Raccoon

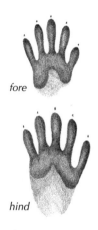

fore

hind

Fore Print
Length: 2–3 in (5–7.5 cm)
Width: 1.8–2.5 in (4.5–6.5 cm)

Hind Print
Length: 2.4–3.8 in (6–9.5 cm)
Width: 2–2.5 in (5–6.5 cm)

Straddle
3.3–6 in (8.5–15 cm)

Stride
Walking: 8–18 in (20–45 cm)
Bounding: 15–25 in (38–65 cm)

Size (female is slightly smaller)
Length: 24–37 in (60–95 cm)

Weight
11–35 lb (5–16 kg)

walking

bounding group

RACCOON
Procyon lotor

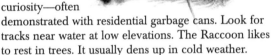

The inquisitive Raccoon is common throughout the region. Adored by some people for its distinctive face mask, it is disliked for its boundless curiosity—often demonstrated with residential garbage cans. Look for tracks near water at low elevations. The Raccoon likes to rest in trees. It usually dens up in cold weather.

The Raccoon's unusual print, showing five well-formed toes, looks like a human handprint; its small claws make dots. Its highly dexterous forefeet rarely leave heel prints, but its hind prints, which are generally much clearer, do show heels. The Raccoon's peculiar walking track pattern shows the left fore print next to the right hind print (or just in front) and vice versa. If a Raccoon is out in deep snow, it may use a direct-registering walk. The Raccoon occasionally bounds, leaving clusters with the two hind prints in front of the fore prints.

Similar Species: Unclear Opossum (p. 40) prints may look similar, but the Opossum drags its tail. In mud or wet snow, River Otter (p. 44) tracks may look similar. Woodchuck (p. 66) prints may also look similar.

Opossum

fore

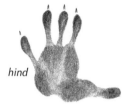

hind

Fore Print
Length: 2–2.3 in (5–5.8 cm)
Width: 2–2.3 in (5–5.8 cm)
Hind Print
Length: 2.5–3 in (6.5–7.5 cm)
Width: 2–3 in (5–7.5 cm)
Straddle
4–5 in (10–13 cm)
Stride
5–11 in (13–28 cm)
Size
Length: 2–2.5 ft (60–75 cm)
Weight
9–13 lb (4–6 kg)

walking   *fast walking*

OPOSSUM
Didelphis virginiana

This slow-moving, nocturnal marsupial is found throughout the region. It occupies many habitats and is quite tolerant of residential areas, but it prefers open woodland or brushland around waterbodies. Look for Opossum tracks in mud near the water and in snow during the warmer months of winter (the Opossum dens up during freezing weather) or near roadkill (which it likes to eat, though many Opossums suffer the same fate as the carrion that they dine on).

The Opossum has two walking habits: the common alternating pattern, with the hind prints registering on the fore prints, and a Raccoon-like (p. 38) paired-print pattern, with each hind print next to the opposing fore print. The very distinctive, long, inward-pointing thumb of the hind foot does not make a claw mark. The Opossum is an excellent climber. In snow, the long, naked tail's dragline may be bloodstained; not well adapted to cold, this thinly haired animal frequently suffers frostbite.

Similar Species: Prints in which the distinctive thumbs do not show may be mistaken for a Raccoon's.

41

Harbor Seal

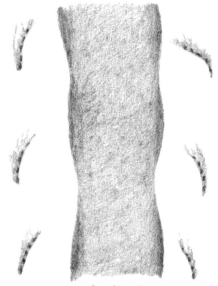

beach tracks

Size (male>female)
Length: 4–6 ft (1.2–1.8 m)
Weight
180–310 lb (80–140 kg)

HARBOR SEAL
Phoca vitulina

This common seal can be found on isolated coastal beaches of this region, especially around Assateague Island. One of the smaller seals, it is quite shy, and it will usually slide off its rocky sentry post into the sea to escape curious naturalists. Look in sandy and muddy areas between platforms for its tracks—unmistakable because of their location and large size. Sometimes a Harbor Seal works its way up a river, so you may find its tracks along riverbanks.

Seals, though unrivaled in the water, are not the most graceful of animals on land. A seal's heavy, fat body and flipper-like feet can leave messy tracks: a wide trough, (made by its cumbersome belly) with dents alongside (made as the seal pushed itself along with its forefeet). Look for the marks made by the seal's nails.

Similar Species: There are no other seals in the region. The unusual shape of the tracks makes confusion with the prints of other animals unlikely.

River Otter

fore

hind

Fore Print
Length: 2.5–3.5 in (6.5–9 cm)
Width: 2–3 in (5–7.5 cm)
Hind Print
Length: 3–4 in (7.5–10 cm)
Width: 2.3–3.3 in (5.8–8.5 cm)
Straddle
4–9 in (10–23 cm)
Stride
Loping: 12–27 in (30–70 cm)
Size
(female is two-thirds the size of male)
Length with tail: 3–4.3 ft (90–130 cm)
Weight
10–25 lb (4.5–11 kg)

loping (fast)

RIVER OTTER
Lontra canadensis

No animal knows how to have more fun than a River Otter. If you are lucky enough to watch one at play, you will not soon forget the experience. Once widespread, this lithe animal has been greatly reduced in numbers by over-trapping and pollution, but in Maryland, Delaware and Virginia its numbers may be increasing. Well-adapted for the aquatic environment, this otter lives along water-bodies; an otter in the forest is usually on its way to another waterbody. The River Otter loves to slide in mud, wet grass or snow, often down riverbanks, leaving troughs nearly 1 foot (30 cm) wide.

In soft mud, the River Otter's five-toed feet, especially the hind ones, register evidence of webbing. The hind foot's inner toe is set slightly apart from the rest. If the forefoot's metacarpal pad registers, it lengthens the print. Very variable, otter trails usually show a typical mustelid 2×2 loping, but with faster gaits they show groups of four and three prints. The thick, heavy tail often leaves a dragline.

Similar Species: Mink (p. 46) and Long-tailed Weasel (p. 48) prints are about half the size, and there will not be a conspicuous tail dragline.

Mink

fore

hind

Fore and Hind Prints
Length: 1.3–2 in (3.3–5 cm)
Width: 1.3–1.8 in (3.3–4.5 cm)

Straddle
2.1–3.5 in (5.3–9 cm)

Stride
Walking/2×2 loping: 8–36 in (20–90 cm)

Size (male>female)
Length with tail: 19–28 in (48–70 cm)

Weight
1.5–3.5 lb (0.7–1.6 kg)

2×2 loping

MINK
Mustela vison

The lustrous Mink, found scattered throughout the region, prefers watery habitats surrounded by brush or forest. At home as much on land as in water, this nocturnal hunter can be exciting to track. Like the River Otter (p. 44), the Mink slides in mud and snow, carving out a trough up to 6 inches (15 cm) wide.

The Mink's fore print shows five (perhaps four) toes, with five loosely connected palm pads in an arc, but the hind print shows only four palm pads. The metacarpal pad of the forefoot rarely registers, but the furred heel of the hind foot may register, lengthening the hind print. The Mink prefers the typical mustelid 2×2 loping gait, which leaves consistently spaced, slightly angled double prints. Its diverse track patterns also include alternating walking, loping with three- and four-print groups (like the River Otter) and bounding (like a rabbit or hare, pp. 56–59).

Similar Species: The Long-tailed Weasel (p. 48) makes similar but generally smaller tracks. The River Otter's trail shows distinct tail draglines. Bobcat (p. 34) prints may resemble four-toed Mink prints, but they will not show claw marks or lobed palm pads.

Weasels

all weasels

Long-tailed Weasel
Fore and Hind Prints
Length: 1.1–1.8 in (2.8–4.5 cm)
Width: 0.8–1 in (2–2.5 cm)
Straddle
1.8–2.8 in (4.5–7 cm)
Stride
Bounding: 9.5–43 in (24–110 cm)
Size (male>female)
Length with tail: 12–22 in (30–55 cm)
Weight
3–9.5 oz (85–270 g)

*Long-tailed
2×2 loping*

LONG-TAILED WEASEL
Mustela frenata

The widely distributed Long-tailed Weasel is an active year-round hunter that has an avid appetite for rodents. Following this nimble creature's tracks can reveal much about its activities. Some weasel trails may lead you up a tree. Weasels sometimes take to water. Tracks are most evident in winter, when weasels frequently burrow into the snow or pursue rodents into their holes. In deep snow, look for holes where the weasel has suddenly plunged in, perhaps in pursuit of prey.

A weasel's light weight, rapid movement and small, hairy feet mean that the print detail is often unclear. Even with good tracks, the inner (fifth) toe rarely registers. The usual weasel gait is a 2×2 lope, leaving a trail of paired prints, but the Long-tailed Weasel's typical 2×2 lope shows an irregular stride—sometimes short and sometimes long—with no consistent behavior. Like the Mink (p. 46), this weasel may bound like a rabbit or hare (pp. 56–59).

Similar Species: Large Short-tailed Weasel (p. 50) tracks or small Mink tracks may be indistinguishable from Long-tailed Weasel tracks. The Least Weasel (p. 50) makes similar but smaller tracks.

Short-tailed Weasel

Fore and Hind Prints
Length: 0.8–1.3 in (2–3.3 cm)
Width: 0.5–0.6 in (1.3–1.5 cm)

Straddle
1–2.1 in (2.5–5.3 cm)

Stride
2x2 loping: 9–36 in (23–90 cm)

Size (male>female)
Length with tail:
   8–14 in (20–35 cm)

Weight
1.8–6 oz (45–170 g)

Least Weasel

Fore and Hind Prints
Length: 0.5–0.8 in (1.3–2 cm)
Width: 0.4–0.5 in (1–1.3 cm)

Straddle
0.8–1.5 in (2–3.8 cm)

Stride
2×2 loping: 5–20 in (13–50 cm)

Size (male>female)
Length with tail:
   6.5–9 in (17–23 cm)

Weight
1.3–1.8 oz (37–51 g)

*Short-tailed
2×2 loping*

*Least
2×2 loping*

SHORT-TAILED WEASEL
(Ermine, Stoat)
Mustela erminea

Not widely distributed, this mid-sized weasel might be found in Maryland, where it prefers woodlands and meadows up to higher elevations, but it does not favor wetlands or dense coniferous forests.

The Short-tailed Weasel's 2×2 loping tracks may fall in clusters, with alternating short and long strides.

Similar Species: A small Long-tailed's (p. 48) tracks may be the same size as a large Short-tailed's. Large Least Weasel (below) tracks may be the same size as a small Short-tailed's.

LEAST WEASEL
Mustela nivalis

The region's smallest weasel makes the least clear tracks—look around wetlands and in open woodlands and fields in western Maryland and western Virginia.

This nimble weasel's trail may be an obstacle course, because it runs over and under logs, up and down trees, and in and out of holes.

Similar Species: Small Short-tailed Weasel (above) tracks can look very similar to large Least Weasel tracks.

Striped Skunk

fore

hind

Fore Print
Length: 1.5–2.2 in (3.8–5.5 cm)
Width: 1–1.5 in (2.5–3.8 cm)

Hind Print
Length: 1.5–2.5 in (3.8–6.5 cm)
Width: 1–1.5 in (2.6–3.8 cm)

Straddle
2.8–4.5 in (7–11 cm)

Stride
Walking/Running:
    2.5–8 in (6.5–20 cm)

Size
Length with tail:
    20–32 in (50–80 cm)

Weight
6–14 lb (2.7–6.5 kg)

walking (fast) | *bounding*

STRIPED SKUNK
Mephitis mephitis

This striking skunk has a notorious reputation for its vile smell, and the lingering odor is often the best sign of its presence. Widespread throughout these states, it prefers lower elevations in a diversity of habitats. The Striped Skunk dens up in winter, coming out on warmer days and in spring.

Both the forefeet and hind feet have five toes. The long claws on the forefeet often register. The smooth palm pads and small heel pads leave surprisingly small prints. The Striped Skunk mostly walks—with such a potent smell for its defense, and those memorable black and white stripes, it rarely needs to run. Note that this skunk's trail rarely shows any consistent pattern, but an alternating walking pattern may be evident. The greater a skunk's speed, the more the hind foot oversteps the fore. If it runs, its trail consists of clumsy, closely set four-print groups. In snow it drags its feet.

Similar Species: The Eastern Spotted Skunk (p. 54) makes smaller prints in a very random pattern. Mink (p. 46) or Long-tailed Weasel (p. 48) tracks will be farther apart than a skunk's. Skunk prints do not overlap.

Eastern Spotted Skunk

fore

hind

Fore Print
Length: 1–1.3 in (2.5–3.3 cm)
Width: 0.9–1.1 in (2.3–2.8 cm)
Hind Print
Length: 1.2–1.5 in (3–3.8 cm)
Width: 0.9–1.1 in (2.3–2.8 cm)
Straddle
2–3 in (5–7.5 cm)
Stride
Walking: 1.5–3 in (3.8–7.5 cm)
Jumping: 6–12 in (15–30 cm)
Size
Length: 13–25 in (33–65 cm)
Weight
0.6–2.2 lb (0.3–1 kg)

walking | bounding

EASTERN SPOTTED SKUNK
Spilogale putorius

This beautifully marked
skunk, which is smaller
than its striped cousin
(p. 52), is found throughout much of Virginia and some
of Maryland. It enjoys diverse habitats—such as scrub-
land, forests and farmland—but it is a rare sight because
of its nocturnal habits, and because it dens up in winter,
coming out only on warmer nights.

This skunk leaves a very haphazard trail as it forages
for food on the ground. Long claws on the forefeet often
register, and the palm and heel may leave defined pad
marks. Although this skunk rarely runs, when it does so
it may bound along, leaving groups of four prints, hind
in front of fore. It occasionally climbs trees, which it does
with ease. It sprays only when truly provoked, so its
powerful odor is less frequently detected than that of the
Striped Skunk.

Similar Species: The Striped Skunk makes larger, less
scattered tracks with a shorter running stride (or it jumps);
it does not climb trees.

Snowshoe Hare

fore

hind

Fore Print
Length: 2–3 in (5–7.5 cm)
Width: 1.5–2 in (3.8–5 cm)
Hind Print
Length: 4–6 in (10–15 cm)
Width: 2–3.5 in (5–9 cm)
Straddle
6–8 in (15–20 cm)
Stride
Hopping: 0.8–4.3 ft (25–130 cm)
Size
Length: 12–21 in (30–53 cm)
Weight
2–4 lb (0.9–1.8 kg)

hopping

SNOWSHOE HARE
(Varying Hare)
Lepus americanus

This hare is well known for its color change from summer brown to winter white and for its huge hind feet, which enable it to 'float' on top of snow. Found in the mountainous areas of western Virginia and western Maryland, it frequents brushy areas and forests, which provide good cover from predators such as the Coyote (p. 26). Hares are most active at twilight and during the night.

The Snowshoe Hare's most common track pattern is a hopping one that shows triangular four-print groups; they can be quite long if the hare was moving quickly. In winter, heavy fur on the hind feet (much larger than the forefeet) thickens the toes, which can splay out to further distribute the hare's weight on snow. Hares make well-worn runways that are often used as escape runs. You may encounter a resting hare, because hares do not live in burrows. Twigs and stems neatly cut at a 45° angle also indicate this hare's presence.

Similar Species: The Eastern Cottontail (p. 58) and the Appalachian (Allegheny) Cottontail (*Sylvilagus obscurus*) make similar but much smaller prints.

Eastern Cottontail

fore

hind

Fore Print
Length: 1–1.5 in (2.5–3.8 cm)
Width: 0.8–1.3 in (2–3.3 cm)

Hind Print
Length: 3–3.5 in (7.5–9 cm)
Width: 1–1.5 in (2.5–3.8 cm)

Straddle
4–5 in (10–13 cm)

Stride
Hopping: 0.6–3 ft (18–90 cm)

Size
Length: 12–17 in (30–43 cm)

Weight
1.3–3 lb (0.6–1.4 kg)

hopping

EASTERN COTTONTAIL
Sylvilagus floridanus

This abundant rabbit is widespread throughout the region. Preferring brushy areas in grasslands and cultivated areas, it might be found in dense vegetation, hiding from predators such as the Bobcat (p. 34) and the Coyote (p. 26). Largely nocturnal, the Eastern Cottontail might be seen at dawn or dusk and on darker days.

As with other rabbits and hares, this rabbit's most common track pattern is a triangular four-print grouping, with the larger hind prints (which can appear pointed) in front of the fore prints (which may overlap). The hairiness of the toes will hide any pad detail. If you follow this rabbit's trail, you could be startled if it flies out from its 'form,' a depression in the ground in which it rests.

Similar Species: The New England Cottontail (*S. transitionalis*) and the Appalachian (Allegheny) Cottontail (*S. obscurus*) make similar tracks, but both are less common, and they are found only in the western part of this region. The Snowshoe Hare (p. 56) makes much larger prints. Squirrel (pp. 70–77) tracks will show the fore prints more consistently side by side.

Nutria

fore

hind

Fore Print
Length: to 3 in (7.5 cm)
Width: to 3 in (7.5 cm)
Hind Print
Length: 4.5–6 in (11–15 cm)
Width: to 3.5 in (9 cm)
Straddle
to 7 in (18 cm)
Stride
Walking: to 8 in (20 cm)
Size (male>female)
Length with tail: 25–55 in (65–140 cm)
Weight
5–25 lb (2.3–11 kg)

walking

NUTRIA
Myocastor coypus

Much larger than a Muskrat (p. 64), this rodent was introduced from South America by fur farmers. It has escaped into the wild and formed numerous colonies in various states, including Virginia and Maryland. With its voracious appetite, the Nutria can be quite destructive to agricultural activities.

The webbing on the Nutria's strong hind feet is often evident in its prints. Look for claw marks, too. The large hind print shows five toes, with one toe set farther back than the others. The much smaller fore print also shows five toes. The large, round, hairless tail often leaves a dragline. A Nutria's trail may lead you to a den in a riverbank, which may be a former Muskrat residence. Nearby you might also notice large mats of vegetation— a Nutria's feeding platform.

Similar Species: Beaver (p. 62) prints are similar, but a Beaver's hind foot has more webbing (extending to the fifth toe). Muskrat prints are smaller.

Beaver

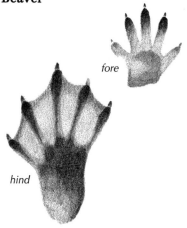

fore

hind

Fore Print
Length: 2.5–4 in (6.5–10 cm)
Width: 2–3.5 in (5–9 cm)
Hind Print
Length: 5–7 in (13–18 cm)
Width: 3.3–5.3 in (8.5–13 cm)
Straddle
6–11 in (15–28 cm)
Stride
Walking: 3–6.5 in (7.5–17 cm)
Size
Length with tail: 3–4 ft (90–120 cm)
Weight
28–75 lb (13–34 kg)

walking

BEAVER
Castor canadensis

Few animals leave as many signs of their presence as the Beaver, which is capable of changing the local landscape. North America's largest rodent, it is a common sight around water. Look for its conspicuous dams and lodges and for the stumps of felled trees. Check trunks gnawed clean of bark for marks of the Beaver's huge incisors. Scent mounds marked with castoreum, a strong-smelling yellowish fluid that Beavers produce, also indicate recent activity.

Check the large hind prints for signs of webbing and broad toenails. The nail of the fourth toe usually does not register, and it is rare for all five toes on each foot to do so. Irregular foot placement in the alternating walking gait may produce a direct register or a double register. The Beaver's thick, scaly tail may mar its tracks, as can the branches that it drags about for construction and food. Repeated path use results in well-worn trails.

Similar Species: The Beaver's many signs and large hind prints minimize confusion. Nutria (p. 60) prints are similar, but there will be no webbing between the fourth and fifth toes. Muskrat (p. 64) prints are much smaller.

Muskrat

fore

hind

Fore Print
Length: 1.1–1.5 in (2.8–3.8 cm)
Width: 1.1–1.5 in (2.8–3.8 cm)
Hind Print
Length: 1.6–3.2 in (4–8 cm)
Width: 1.5–2.1 in (3.8–5.3 cm)
Straddle
3–5 in (7.5–13 cm)
Stride
Walking: 3–5 in (7.5–13 cm)
Running: to 1 ft (30 cm)
Size
Length with tail: 16–25 in (40–65 cm)
Weight
2–4 lb (0.9–1.8 kg)

walking

MUSKRAT
Ondatra zibethicus

Like the Beaver (p. 62), this rodent is found throughout the region, wherever there is water. Beavers are very tolerant of Muskrats and even allow them to live in parts of their lodges. Active all year, the Muskrat leaves plenty of signs. It digs extensive networks of burrows, often undermining riverbanks, so do not be surprised if you suddenly fall into a hidden hole! Also look for small lodges in the water and for the beds of vegetation on which the Muskrat rests, suns and feeds in summer.

The small fifth (innermost) toe of the forefoot rarely registers. Stiff hairs that aid in swimming may create a 'shelf' around the five well-formed toes of the hind print. The common alternating walking pattern shows print pairs that alternate from side to side; the hind print is just behind the fore print or slightly overlaps it. In snow, a Muskrat's feet drag, and its tail leaves a sweeping dragline.

Similar Species: Few animals share this water-loving rodent's habits. The Beaver makes larger tracks.

Woodchuck

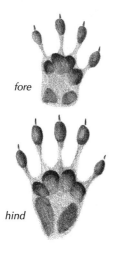

fore

hind

Fore and Hind Prints
Length: 1.8–2.8 in (4.5–7 cm)
Width: 1–2 in (2.5–5 cm)

Straddle
3.3–6 in (8.5–15 cm)

Stride
Walking: 2–6 in (5–15 cm)
Bounding: 6–14 in (15–35 cm)

Size (male>female)
Length with tail:
    20–25 in (50–65 cm)

Weight
5.5–12 lb (2.5–5.5 kg)

walking   *bounding*

WOODCHUCK
(Whistle Pig, Groundhog, Marmot)
Marmota monax

This robust member of the squirrel family is a common sight in open woodlands and adjacent open areas throughout this region, except in the very south of Virginia. Always on the watch for predators, but not too troubled by humans, the Woodchuck never wanders far from its burrow. This marmot hibernates during winter but emerges in early spring; look for tracks in late spring snowfalls and in mud around the burrow entrances.

A Woodchuck's fore print shows four toes, three palm pads and two heel pads (not always evident). The hind print shows five toes, four palm pads and two poorly registering heel pads. The Woodchuck usually leaves an alternating walking pattern, with the hind print registered on the fore print. When a Woodchuck runs from danger, it makes groups of four prints, hind ahead of fore.

Similar Species: Squirrel (pp. 70–77) prints are very similar but smaller. A small Raccoon's (p. 38) bounding track pattern will be similar, but it will show five-toed fore prints.

Eastern Chipmunk

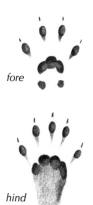

fore

hind

Fore Print
Length: 0.8–1 in (2–2.5 cm)
Width: 0.4–0.8 in (1–2 cm)

Hind Print
Length: 0.7–1.3 in (1.8–3.3 cm)
Width: 0.5–0.9 in (1.3–2.3 cm)

Straddle
2–3.2 in (5–8 cm)

Stride
Running: 7–15 in (18–38 cm)

Size
Length with tail: 7–10 in (18–25 cm)

Weight
2.5–5 oz (70–140 g)

bounding

EASTERN CHIPMUNK
Tamias striatus

This delightful character lives throughout most of the region, except for southern Virginia. The Eastern Chipmunk is found in a variety of habitats, from the dense forest floor to open areas near buildings. You are more likely to see or hear this animal, which is highly active during summer, than to notice its tracks. This large chipmunk is happiest on the ground, but it will gladly climb sturdy oak trees to harvest juicy, ripe acorns. It hibernates in winter, waking up from time to time to eat.

Chipmunks are so light that their tracks rarely show fine details. The forefeet each have four toes and the hind feet have five. Chipmunks run on their toes, so the two heel pads of the forefeet seldom register; the hind feet have no heel pads. Their erratic track patterns, like those of many of their cousins, show the hind feet registered in front of the forefeet. A chipmunk trail often leads to extensive burrows.

Similar Species: No other chipmunks live in this region. Most squirrels (pp. 70–77) have larger prints and a wider straddle, and they are more likely to make midwinter tracks. Mouse (pp. 86–89) tracks are smaller.

Eastern Gray Squirrel

fore

hind

Fore Print
Length: 1–1.8 in (2.5–4.5 cm)
Width: 1 in (2.5 cm)
Hind Print
Length: 2.3–3 in (5.8–7.5 cm)
Width: 1.1–1.5 in (2.8–3.8 cm)
Straddle
3.8–6 in (9.5–15 cm)
Stride
Bounding: 0.7–3 ft (22–90 cm)
Size
Length with tail: 17–20 in (43–50 cm)
Weight
14–25 oz (400–710 g)

bounding

EASTERN GRAY SQUIRREL

Sciurus carolinensis

This large, familiar squirrel can be a common sight in deciduous and mixed forests throughout this region, even in urban areas. Active all year, the Eastern Gray Squirrel can leave a wealth of evidence, especially in winter as it scurries about digging up nuts that it buried during the previous fall.

The Eastern Gray Squirrel leaves a typical squirrel track pattern when it runs or bounds. The hind prints fall slightly in front of the fore prints. A clear fore print shows four toes with sharp claws, four fused palm pads and two heel pads. The hind print shows five toes and four palm pads; if the full heel-length registers, it also shows two small heel pads.

Similar Species: Fox Squirrel (p. 72) prints are as big or larger. Red Squirrel (p. 74) prints are smaller. Chipmunks (p. 68) and flying squirrels (p. 76) make smaller tracks in a similar pattern, but they have narrower straddles. Rabbits and hares (pp. 56–59) make longer track patterns; their forefeet rarely register side by side when they run.

71

Fox Squirrel

fore

hind

Fore Print
Length: 1–1.9 in (2.5–4.5 cm)
Width: 1–1.7 in (2.5–4.3 cm)
Hind Print
Length: 2–3.3 in (5–7.5 cm)
Width: 1.5–1.9 in (3.8–4.8 cm)
Straddle
4–6 in (10–15 cm)
Stride
Bounding: 0.7–3 ft (22–90 cm)
Size
Length with tail: 18–28 in (45–70 cm)
Weight
1–2.4 lb (0.5–1.1 kg)

bounding

FOX SQUIRREL
Sciurus niger

This squirrel is much like the Eastern Gray Squirrel (p. 70), but it is larger and has a yellowish underside. It can be a common sight in deciduous forests with plenty of nut trees and in open areas or woodlands throughout these states. Piles of nutshells at tree bases indicate its favorite feeding sites. Active all year, the Fox Squirrel spends a lot of time foraging on the ground, often collecting nuts that it buried singly during the previous fall. A well-known locally endemic subspecies is the Delmarva Squirrel.

A clear fore print shows four toes with claws evident, four fused palm pads and two heel pads. The hind print shows five toes, four palm pads and sometimes a heel. When this squirrel runs or bounds, its typical squirrel pattern shows the hind prints slightly in front of the fore prints and the prints in each pair roughly side by side.

Similar Species: Eastern Gray Squirrel prints are generally slightly smaller. Chipmunk (p. 68), Red Squirrel (p. 74) and flying squirrel (p. 76) tracks fall in a similar pattern, but they are smaller, and the straddles are narrower. Rabbit and hare (pp. 56–59) bounding groups are longer, and the fore prints rarely register side by side.

Red Squirrel

fore

hind

Fore Print
Length: 0.8–1.5 in (2–3.8 cm)
Width: 0.5–1 in (1.3–2.5 cm)

Hind Print
Length: 1.5–2.3 in (3.8–5.8 cm)
Width: 0.8–1.3 in (2–3.3 cm)

Straddle
3–4.5 in (7.5–11 cm)

Stride
Bounding: 8–30 in (20–75 cm)

Size
Length with tail:
9–15 in (23–38 cm)

Weight
2–9 oz (57–260 g)

bounding

*bounding
(deep snow)*

RED SQUIRREL
(Pine Squirrel, Chickaree)
Tamiasciurus hudsonicus

When you
enter a Red
Squirrel's terri-
tory, the inhabi-
tant greets you
with a loud, chat-
tering call. Another
obvious sign of this forest
dweller, which is found in
much of the region (but not southern Virginia), is its
large middens—piles of cone scales and cores left
beneath trees—that mark favorite feeding sites.

Active all year in its small territory, a Red Squirrel
will leave an abundance of trails that lead from tree
to tree or down a burrow. This energetic animal mostly
bounds, leaving groups of four prints, the hind prints
in front of the fore prints, which tend to be side by side
(but not always). Four toes show on each fore print, and
five show on each hind print. The heels often do not
register when squirrels move quickly. In deeper snow
the prints merge to form pairs of diamond-shaped tracks.

Similar Species: The larger Fox Squirrel (p. 72) and
Eastern Gray Squirrel (p. 70) both make similar but
larger tracks. Eastern Chipmunk (p. 68) and Southern
Flying Squirrel (p. 76) tracks fall in a similar pattern,
but they are smaller and have narrower straddles.

Southern Flying Squirrel

fore

hind

Fore Print
Length: 0.3–0.5 in (0.8–1.3 cm)
Width: 0.4 in (1 cm)
Hind Print
Length: 0.9–1.3 in (2.3–3.3 cm)
Width: 0.5 in (1.4 cm)
Straddle
2–2.5 in (5–6.5 cm)
Stride
Bounding: 7–22 in (18–55 cm)
Size
Length with tail: 8–10 in (20–25 cm)
Weight
1.5–3.2 oz (43–90 g)

walking

SOUTHERN FLYING SQUIRREL

Glaucomys volans

This soft-furred brown acrobat is capable of long-distance gliding using the membranes that connect its forelegs and hind legs. Mainly nocturnal, the Southern Flying Squirrel lives in coniferous and mixed forests throughout the region. Up to 50 of these squirrels may huddle together in a nest for warmth in winter, but they do not truly hibernate.

When it is on the ground, this flying squirrel normally bounds. Although its hind prints may register in front of its fore prints, New England-based tracking expert Mark Elbroch reports that the more common pattern (with a narrower straddle) shows the fore prints in front. If it glides down to the ground, this squirrel can leave a distinctive four-print 'sitzmark' in loose material or snow, but it often climbs down a tree trunk instead.

Similar Species: Other squirrel (pp. 70–75) prints will usually be larger, but tracks in loose dirt or snow can be difficult to identify. Chipmunks (p. 68) have smaller tracks with a narrower straddle.

Appalachian Woodrat

fore

hind

Fore Print
Length: 0.6–0.8 in (1.5–2 cm)
Width: 0.4–0.5 in (1–1.3 cm)

Hind Print
Length: 1–1.5 in (2.5–3.8 cm)
Width: 0.6–0.8 in (1.5–2 cm)

Straddle
2.3–2.8 in (5.8–7 cm)

Stride
Walking: 1.8–3 in (4.5–7.5 cm)
Bounding: 5–8 in (13–20 cm)

Size
Length with tail:
    14–17 in (35–43 cm)

Weight
13–16 oz (370–450 g)

walking   *bounding*

APPALACHIAN WOODRAT
(Allegheny Woodrat)
Neotoma magister

   This nocturnal woodrat thrives in rocky areas throughout most of this region, except southeastern Virginia. Its trail might lead you to a distinctive mass of a nest, most often under a rock or in a crevice. Although the Appalachian Woodrat favors rocky areas, it feeds on foliage, seeds, ferns and fungi; therefore, it requires nearby vegetation.

   Four toes show on the fore print and five on the hind. The short claws rarely register. A woodrat often walks in an alternating fashion, with the hind print direct registering on the fore print. This woodrat frequently bounds as well, leaving a pattern of four prints, with the larger hind print in front of the diagonally placed fore prints. The stride tends to be short relative to the size of the prints.

Similar Species: Indistinct squirrel (pp. 70–77) prints are similar but usually larger. The Norway Rat (p. 80) has similar prints, but it is usually found close to human activity. Woodchuck (p. 66) prints are similar but much larger.

Norway Rat

fore

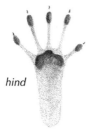

hind

Fore Print
Length: 0.7–0.8 in (1.8–2 cm)
Width: 0.5–0.7 in (1.3–1.8 cm)

Hind Print
Length: 1–1.3 in (2.5–3.3 cm)
Width: 0.8–1 in (2–2.5 cm)

Straddle
2–3 in (5–7.5 cm)

Stride
Walking: 1.5–3.5 in (3.8–9 cm)
Bounding: 9–20 in (23–50 cm)

Size
Length with tail: 13–19 in (33–48 cm)

Weight
7–18 oz (200–510 g)

walking

NORWAY RAT
(Brown Rat)
Rattus norvegicus

Active both day
and night, this despised rat
is widespread almost anywhere
that humans have decided to build
their homes. Not entirely dependent
on people, it may live in the wild as well.

The fore print shows four toes, and the hind print
shows five. When it bounds, this colonial rat leaves four-
print groups, with the hind prints in front of the diagonally placed fore prints. Sometimes one of the hind feet
direct registers on a fore print, creating a three-print
group. This rat more commonly leaves an alternating
walking pattern with the larger hind prints close to or
overlapping the fore prints; the hind heel does not show.
The tail often leaves a dragline in loose material. Rats
live in groups, so you may find many trails together,
often leading to their 5-inch (2-cm) wide burrows.

Similar Species: The Black Rat (*R. rattus*) and the
Hispid Cotton Rat (p. 82), both also widespread, leave
similar tracks. The Appalachian Woodrat (p. 78) leaves
similar tracks, but it seldom associates with human activity. Mouse (pp. 86–89) prints are much smaller. Squirrel
(pp. 70–77) tracks show distinctive squirrel traits. The
Eastern Chipmunk (p. 68) usually bounds, and its tracks
are often smaller.

Hispid Cotton Rat

fore

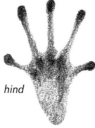

hind

Fore Print
Length: 0.5–0.7 in (1.3–1.8 cm)
Width: 0.5–0.7 in (1.3–1.8 cm)

Hind Print
Length: 0.6–1 in (1.5–2.5 cm)
Width: 0.6–0.8 in (1.5–2 cm)

Straddle
1.3–1.5 in (3.3–3.8 cm)

Stride
Walking: 1.3 in (3.3 cm)

Size
Length with tail: 8–14 in (20–35 cm)

Weight
2.8–7 oz (80–200 g)

walking

HISPID COTTON RAT
Sigmodon hispidus

The Hispid Cotton Rat, found throughout most of Virginia, makes itself unpopular by eating valuable crops. Although keen on devouring almost anything green, it prefers grassy fields. It stays close to home, and consequently the rat's little runways clearly mark its routes to favored feeding sites.

The fore prints show four toes, but the larger hind prints usually show five. The heel of the hind foot will not always register, especially if the rat is moving fast. This medium-sized rodent leaves a typical walking track pattern in which the hind print registers on and slightly behind the fore print. Watch for its nests, which are balls of woven grass, and small piles of cut grass.

Similar Species: The Norway Rat (p. 80) and the Black Rat (*Rattus rattus*) also make similar tracks, but they are usually found close to human activity.

Woodland Vole

fore

hind

Fore Print
Length: 0.5 in (1.3 cm)
Width: 0.5 in (1.3 cm)

Hind Print
Length: 0.6 in (1.5 cm)
Width: 0.5–0.8 in (1.3–2 cm)

Straddle
1.3–2 in (3.3–5 cm)

Stride
Walking/Trotting: 0.8 in (2 cm)
Bounding: 2–6 in (5–15 cm)

Size
Length with tail:
    4–5.5 in (10–14 cm)

Weight
0.8–1.3 oz (23–37 g)

walking

bounding (in snow)

WOODLAND VOLE
(Pine Vole)

Microtus pinetorum

Distinguishing
a vole track from
the tracks of other
small mammals in the
region can be challenging.
If you do spot a vole track, the
most likely candidate throughout these states is the
Woodland Vole. It lives in a variety of habitats.

When clear (which is seldom), vole fore prints show
four toes, and hind prints show five. A vole's walk and
trot both leave a paired alternating track pattern with a
hind print occasionally direct registered on a fore print.
Voles usually opt for a faster bounding in which the hind
prints register on the fore prints to form print pairs. This
vole lopes quickly across open areas, creating a three-
print track pattern. Voles stay under the snow in winter;
when the snow melts, look for distinctive piles of cut grass
from their ground nests. The bark at the bases of shrubs
may show tiny teeth marks left by gnawing. In summer,
well-used vole paths appear as little runways in the grass.

Similar Species: The Southern Red-backed Vole (*Clethe-
rionomys gapperi*) and the Meadow Vole (*M. pennsylvani-
cus*) are both common and leave indistinguishable tracks.
Mouse (pp. 86–89) bounding groups show four prints.

White-footed Mouse

bounding group

Fore Print
Length: 0.3–0.4 in (0.8–1 cm)
Width: 0.3–0.4 in (0.8–1 cm)

Hind Print
Length: 0.3–0.5 in (0.8–1.3 cm)
Width: 0.3–0.4 in (0.8–1 cm)

Straddle
1.4–1.8 in (3.5–4.5 cm)

Stride
Bounding: 3–12 in (7.5–30 cm)

Size
Length with tail:
6–12 in (15–30 cm)

Weight
0.5–1.3 oz (14–35 g)

bounding

bounding (in snow)

WHITE-FOOTED MOUSE

Peromyscus leucopus

The highly adaptable
White-footed Mouse—one of the
region's most abundant mammals—lives anywhere from
woodlands to cultivated fields. It is seldom seen, because
it is nocturnal. This mouse may enter buildings in winter,
where it will stay active. In colder areas it may hibernate.

In perfect soft mud, the fore prints each show four
toes, three palm pads and two heel pads, and the hind
prints show five toes and three palm pads; the hind heel
pads rarely register. Bounding tracks, most noticeable
in snow, show the hind prints falling in front of the fore
prints. In flaky snow the prints may merge to look like
larger pairs of prints; tail drag will be evident. This
mouse's trail may lead up a tree or down into a burrow.

Similar Species: Other less common mice make indistinguishable prints. The House Mouse (*Mus musculus*),
with similar tracks, associates more with humans. Jumping mouse (p. 88) prints show long, thin toes. Voles
(p. 84) tend to trot, and they have a much shorter bounding track pattern. Chipmunks (p. 68) have a wider straddle. Shrews (p. 90) have a narrower straddle.

Meadow Jumping Mouse

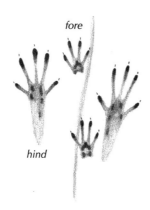

fore

hind

Fore Print
Length: 0.3–0.5 in (0.8–1.3 cm)
Width: 0.3–0.5 in (0.8–1.3 cm)
Hind Print
Length: 0.5–1.3 in (1.3–3.3 cm)
Width: 0.5–0.7 in (1.3–1.8 cm)
Straddle
1.8–1.9 in (4.5–4.8 cm)
Stride
Bounding: 7–18 in (18–45 cm)
In alarm: 3–6 ft (90–180 cm)
Size
Length with tail: 7–9 in (18–23 cm)
Weight
0.6–1.3 oz (17–35 g)

bounding

MEADOW JUMPING MOUSE

Zapus hudsonius

Congratulations if you find and successfully identify the tracks of the Meadow Jumping Mouse! Although it can be locally abundant, its preference for grassy meadows and its long, deep winter hibernation (about six months!) make locating tracks very difficult.

Jumping mouse tracks are distinctive if you do find them. The two smaller fore prints are between the long hind prints; the long heels do not always register, and some prints show just the three long middle toes. The toes on the forefeet may splay so much that the side toes point backward. When they bound, jumping mice make short leaps. The tail may leave a dragline in soft mud or unseasonable snow. Clusters of cut grass stems, about 5 inches (13 cm) long, found lying in meadows are a more abundant sign of this rodent.

Similar Species: Woodland Jumping Mouse (*Napaeozapus insignis*) tracks are similar; they can be found in western Maryland and western Virginia. Heel-less jumping mouse hind prints may resemble a vole's (p. 84), a small bird's (p. 112) or an amphibian's (pp. 114–19).

Masked Shrew

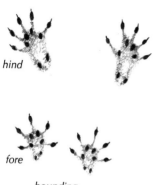

hind

fore

bounding group

Fore Print
Length: 0.2 in (0.5 cm)
Width: 0.2 in (0.5 cm)

Hind Print
Length: 0.6 in (1.5 cm)
Width: 0.3 in (0.8 cm)

Straddle
0.8–1.3 in (2–3.3 cm)

Stride
Bounding: 1.2–2.3 in (3–5.8 cm)

Size
Length with tail: 3–4 in (7.5–10 cm)

Weight
0.1–0.2 oz (3–6 g)

bounding

MASKED SHREW
(Cinereus Shrew)

Sorex cinereus

Though several species of tiny, frenetic shrews
live in this region, the abundant Masked Shrew
is a likely candidate if you find tracks. This shrew
prefers moist fields, bogs, marshes or woodlands, but
it can also be found in higher and drier grasslands. Its
rapid activity makes it difficult to observe closely.

In its energetic and unending quest for food, a shrew
usually leaves a four-print bounding pattern, but it may
slow to an alternating walking pattern. The individual
prints in a group are often indistinct but, in mud or shallow, wet snow, you can even count the five toes on each
print. In deeper snow, a shrew's tail often leaves a dragline. If a shrew tunnels under snow, it may leave a snow
ridge on the surface. A shrew's trail may disappear down
a burrow.

Similar Species: Other shrews with indistinguishable
tracks in this region include the Smoky Shrew (*S. fumeus*),
the Southeastern Shrew (*S. longirostris*), the Pygmy Shrew
(*S. hoyi*) and the Least Shrew (*Cryptotis parva*). Mouse
(pp. 86–89) fore prints show four toes.

Eastern Mole

a molehill of the Eastern Mole

some molehills and ridges of the Eastern Mole

Size
Length: 4.5–6.5 in (11–17 cm)
Tail length: 1–1.5 in (2.5–3.8 cm)
Weight
2.5–5 oz (70–140 g)

EASTERN MOLE
Scalopus aquaticus

This soft-furred resident of the underworld is the mole that you will most likely encounter in this region. It can certainly leave a wealth of evidence that indicates its presence, usually in pastures or open woodlands, especially where the soil is light and moist and easy to burrow in.

Moles, which seldom emerge from their subterranean environment, create an extensive network of burrows through which they forage. These burrows are sometimes marked by ridges on the surface, though most of us are more familiar with the hills that form where the mole gets rid of excess soil from its burrows and for which moles are frequently considered to be pests (when they mess up fine lawns). When rain moistens the soil and brings worms to the surface, it can be entertaining to watch the earth twitch and rise up as the mole satisfies its voracious appetite.

Similar Species: The Hairy-tailed Mole (*Parascalops breweri*), which is found in Maryland, Delaware and all but central and eastern Virginia, leaves similar signs.

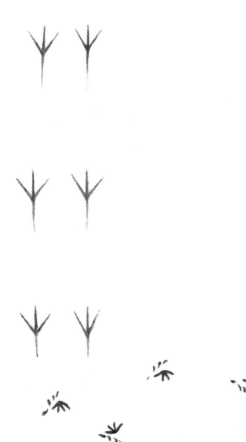

BIRDS, AMPHIBIANS & REPTILES

A guide to the animal tracks of Virginia, Maryland and Delaware is not complete without considering birds, amphibians and reptiles.

Several bird species have been chosen to represent the main types common to this region, but remember that individual bird species are not easily identified by track alone. The shores of lakes and streams are very reliable places to find bird tracks—the mud there can hold a clear print for a long time. The sheer number of tracks made by shorebirds and waterfowl can be astonishing. Bird tracks are also abundant in snow and are clearest in shallow, wet snow. Although some bird species prefer to perch in trees or soar across the sky, it can be entertaining to follow the tracks of birds that spend a lot of time on the ground. They can spin around in circles and lead you in all directions. The trail may suddenly end as the bird takes flight, or it might terminate in a pile of feathers, the bird having fallen victim to a predator.

Many amphibians and turtles depend on moist environments, so look in the soft mud along the shores of lakes and ponds for their distinctive tracks. You may be able to distinguish frog tracks from toad tracks, because these two amphibians generally move differently, but it can be very difficult to identify the species. Reptiles thrive and outnumber the amphibians in drier environments, but they seldom leave good tracks, except in occasional mud or perhaps in sand. Snakes leave distinctive body prints.

Mallard

Print
Length: 2–2.5 in (5–6.5 cm)
Straddle
4 in (10 cm)
Stride
to 4 in (10 cm)
Size
23 in (58 cm)

MALLARD
Anas platyrhynchos

male

female

This dabbling duck—the male a familiar sight with its striking green head—is common in open areas near lakes and ponds. Its webbed feet leave prints that can often be seen in abundance along the muddy shores of just about any waterbody, including those in urban parks.

The webbed foot of the Mallard has three long toes that all point forward. Although the toes register well, the webbing between the toes does not always show in the print. The inward-pointing feet give the Mallard a pigeon-toed appearance, which perhaps accounts for its waddling gait, a characteristic for which ducks are known.

Similar Species: Many waterfowl, such as other ducks, as well as the Herring Gull (p. 98), leave similar prints. Exceptionally large prints were likely made by a goose (various species) or a swan (*Cygnus* spp.).

Herring Gull

Print
Length: 3.5 in (9 cm)
Straddle
4–6 in (10–15 cm)
Stride
4.5 in (11 cm)
Size
Length: 23–25 in (58–65 cm)

HERRING GULL

Larus argentatus

The Herring Gull, with its long wings and webbed toes, is a strong long-distance flier as well as an excellent swimmer. Common in a variety of habitats, it can be found concentrated in great numbers along waterbodies.

Gulls leave slightly asymmetrical tracks that show three toes. They have claws that register outside the webbing, and the claw marks are usually attached to the footprint. Most gulls have quite a swagger to their gait, and they leave a trail with the tracks turned strongly inward.

Similar Species: Gull species cannot be reliably identified by track alone, but smaller species have conspicuously smaller tracks. Mallard (p. 96) and other duck tracks are often difficult to distinguish from gull tracks. Geese (various species) make larger prints.

Great Blue Heron

Print
Length: to 6.5 in (17 cm)
Straddle
8 in (20 cm)
Stride
9 in (23 cm)
Size
4.2–4.5 ft (1.3–1.4 m)

GREAT BLUE HERON
Ardea herodias

The refined and graceful image of this large heron symbolizes the precious wetlands in which it patiently hunts for food. Usually still and statuesque as it waits for a meal to swim by, the Great Blue Heron will have cause to walk from time to time, perhaps to find a better hunting location. Look for its large, slender tracks along the banks or mudflats of waterbodies.

Not surprisingly, a bird that lives and hunts with such precision walks in a similar fashion, leaving straight tracks that fall in a nearly straight line. Look for the slender rear toe in the print.

Similar Species: Other herons and egrets—such as the Green Heron (*Butorides virescens*), the Great Egret (*Casmerodius albus*) and the night herons (*Nycticorax* spp.)—have similar prints that vary in size relative to the size of the bird.

Common Snipe

Print
Length: 1.5 in (3.8 cm)
Straddle
to 1.8 in (4.5 cm)
Stride
to 1.3 in (3.3 cm)
Size
11–12 in (28–30 cm)

COMMON SNIPE
Gallinago gallinago

   This short-legged character is a resident of marshes and bogs, where its neat prints can often be seen in mud. Snipes are quite secretive when on the ground, and so you may be surprised if one suddenly flushes out from beneath your feet. If there is a Common Snipe in the air, you may hear an eerie whistle if it dives from the sky.

   The Common Snipe's neat prints show four toes, including a small rear toe that points inward. The bird's short legs and stocky body give it a very short stride.

Similar Species: Many shorebirds, including the Spotted Sandpiper (p. 104), leave similar tracks.

Spotted Sandpiper

Print
Length: 0.8–1.3 in (2–3.3 cm)
Straddle
to 1.5 in (3.8 cm)
Stride
Erratic
Size
7–8 in (18–20 cm)

SPOTTED SANDPIPER

Actitis macularia

The bobbing tail of the Spotted Sandpiper is a common sight on the shores of lakes, rivers and streams, but you will usually find just one of these territorial birds in any given location. Because of its excellent camouflage, likely the first that you will see of this bird will be when it flies away, its fluttering wings close to the surface of the water.

As a sandpiper teeters up and down on the shore, it leaves trails of three-toed prints that show a very small fourth toe facing off to one side at an angle. Sandpiper tracks can have an erratic stride.

Similar Species: All sandpipers and plovers, such as the common Killdeer (*Charadrius vociferus*), leave similar tracks, although there is much diversity in size. The Common Snipe (p. 102) makes similar but larger tracks.

Great Horned Owl

Strike
Width: to 3 ft (90 cm)
Size
22 in (55 cm)

GREAT HORNED OWL
Bubo virginianus

Often seen resting quietly in trees by day, this wide-ranging owl prefers to hunt at night. Signs of this bird are more common in winter, when the owl often strikes through the snow with its talons, leaving an untidy hole that may be surrounded by wing and tail-feather imprints. If it registers well, this 'strike' can be quite a sight. The feather imprints are made as the owl struggles to take off with possibly heavy prey. An ungraceful walker, it prefers to fly away from the scene.

You may stumble across a strike and guess that the owl's target could have been a vole (p. 84) scurrying around underneath the snow. Or you may be following the surface trail of an animal to find that it abruptly ends with this strike mark where the animal has been seized.

Similar Species: If the prey left no approaching trail (meaning that it was moving under the snow), the strike mark is likely an owl's, because owls hunt by sound. If there is a trail, the strike mark is probably from a bird of prey that hunts by sight instead.

American Crow

Print
Length: 2.5–3 in (6.5–7.5 cm)
Straddle
1.5–3 in (3.8–7.5 cm)
Stride
Walking: 4 in (10 cm)
Size
16 in (40 cm)

108

AMERICAN CROW

Corvus brachyrhyncos

The black silhouette of the American Crow can be a common sight in a variety of habitats. This bird will frequently come down to the ground and contentedly strut around with a confidence that hints at its intelligence. Its loud *caw* can be heard from quite a distance. Crows can be especially noisy when they are mobbing an owl or a hawk.

The American Crow typically leaves an alternating walking track pattern. Its prints show three sturdy toes pointing forward and one toe pointing backward. When a crow is in need of greater speed, perhaps for take-off, it bounds along, leaving irregular pairs of diagonally placed prints with a longer stride between each pair.

Similar Species: Other corvids, such as the Blue Jay (*Cyanocitta cristata*), also spend a lot of time on the ground and make similar tracks that vary in size depending on the size of the bird.

Northern Flicker

Print
Length: 1.8 in (4.5 cm)
Straddle
1–1.5 in (2.5–3.8 cm)
Stride
Hopping: 1.5–5 in (3.8–13 cm)
Size
5.5–6.5 in (14–17 cm)

NORTHERN FLICKER

Colaptes auratus

male

female

This attractive woodpecker, which can be seen throughout the region, is common in open woodlands and right into suburban areas. Unlike most woodpeckers, it spends some of its time feeding on the ground.

A clear flicker track shows a distinctive arrangement of two strong toes pointing forward and two pointing to the rear, with the outer toes slightly longer than the inner ones. The flicker's toes—along with short, strong legs that give the bird a short stride—are well suited for grasping tree trunks and limbs as this agile bird works its way along in search of insects.

Similar Species: Most other birds have very different tracks. Other woodpeckers would leave similar tracks, but very few come down to the ground as much as the obliging Northern Flicker does.

Dark-eyed Junco

Print
Length: to 1.5 in (3.8 cm)
Straddle
1–1.5 in (2.5–3.8 cm)
Stride
Hopping: 1.5–5 in (3.8–13 cm)
Size
5.5–6.5 in (14–17 cm)

DARK-EYED JUNCO
Junco hyemalis

This common small bird typifies the many small hopping birds found in the region. A good place to study this type of prints is near a birdfeeder. Watch the birds scurry around as they pick up fallen seeds, then have a look at the prints left behind. For example, the Dark-eyed Junco is attracted to seeds that chickadees (*Poecile* spp.) scatter as they forage for sunflower seeds in the birdfeeder. Also look for tracks under coniferous trees, where juncos feed on fallen seeds in winter.

Each foot has three forward-pointing toes and one longer rear toe. The best prints are left in mud or snow, although in sand or loose snow the toe detail is lost. The prints may show some dragging between the hops.

Similar Species: Many small birds, such as the Northern Cardinal (*Cardinalis cardinalis*), finches (various species) and sparrows (various species), make similar tracks. Toe size may help with identification—larger birds make larger prints—as can the season. In powdery snow, mouse (pp. 86–89) tracks might appear similar, so follow the trail to see if it disappears down a hole or into thin air.

Frogs

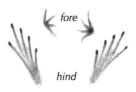

fore

hind

Straddle
to 3 in (7.5 cm)

hopping

FROGS

Bullfrog

The best place to look for frog tracks is along the muddy fringes of waterbodies. A frog generally hops, resulting in its two small forefeet registering in front of its long-toed hind prints. Frog tracks vary greatly in size, depending on species and age.

Of the many frogs in the region, the treefrogs are among the smallest. The Spring Peeper's (*Pseudacris crucifer*) 1.5-inch (3.8-cm) length and its preference for thick undergrowth and shrubs near the water make its tracks a rare sight. The larger gray treefrogs (*Hyla chrysoscelis* and *H. versicolor*) spend most of their time in trees, coming down to breed and sing at night. The Green Frog (*Rana clamitans*) and the Pickerel Frog (*R. palustris*) are both widespread and favor slow-moving, shallow water and swampy areas. The beautiful Southern Leopard Frog (*R. sphenocephala*), which grows to 5 inches (13 cm) long and lives in eastern Virginia, eastern Maryland and all of Delaware, makes larger tracks. Very large tracks are surely from the robust Bullfrog (*R. catesbeiana*). To 8 inches (20 cm) in length, it is North America's largest frog.

Toads

hind *fore*

Straddle
to 2.5 in (6.5 cm)

walking

TOADS

American Toad

Undoubtedly, the best place to look for toad tracks is, as with frog tracks, along the muddy fringes of waterbodies, but they can occasionally be found in drier areas—for example, as unclear trails in dusty patches of soil. Toads generally walk, but they are pretty capable hoppers too, especially when being hassled by overly enthusiastic naturalists. Toads leave rather abstract prints as they walk. The heels of the hind feet do not register. On less firm surfaces, the toes often leave draglines.

There are fewer toad species than frog species in this region. The toad most likely to be encountered, and the most widespread, is the American Toad (*Bufo americanus*); it lives in many different moist habitats. Fowler's Toad (*B. fowleri*) is found scattered throughout Maryland and Virginia in temporary pools and ditches. The Eastern Spadefoot (*Scaphiopus holbrookii*) might be encountered in forests, brushy areas or cultivated land. Toads in this region can grow up to 4.5 inches (11 cm) long.

117

Salamanders

fore

hind

Straddle
to 4 in (10 cm)

walking

SALAMANDERS & NEWTS

Eastern Newt

There are several varieties of salamanders and newts in the moist and wet areas of this region. In general, a salamander fore print shows four toes, and the larger hind print shows five. However, print detail is often blurred by the animal's dragging belly or by the swinging of its thick tail across the tracks. Among the more abundant and widespread of these long, slender, lizard-like amphibians is the Eastern Newt (*Notophthalmus viridescens*), which can grow to 5.5 inches (14 cm) long. After a fresh rain, Eastern Newts emerging from ponds may leave small trails in the mud.

The king of the salamander world is undoubtedly the magnificent Eastern Tiger Salamander (*Ambystoma tigrinum*), which can grow up to 13 inches (33 cm) in length and comes in such a diversity of colors and patterns that it defies description. Found near the coast, this heavy salamander leaves the best tracks. Several other *Ambystoma* species leave tracks of varying sizes, depending on the size of the salamander. The slightly smaller Spotted Salamander (*A. maculatum*) lives in mixed forests and some coniferous forests. Smaller still, the Jefferson Salamander (*A. jeffersonianum*), with its handsome blue spots, is locally common in moist mountain woodlands.

Lizards

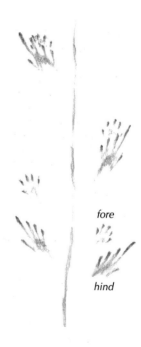

fore

hind

Straddle
to 3 in (7.5 cm)

walking

LIZARDS & SKINKS

Five-lined Skink

Lizards can be difficult to find in this region, but in some woodlands and certain arid areas they can be locally abundant. Their tracks are similar to salamander tracks, but they show longer, more slender toes. These reptiles all move very quickly when the need arises, their feet barely touching the ground as they dart for cover. Consequently, their tracks can be hard to make out clearly.

If you find lizard tracks, a likely candidate is the Five-lined Skink (*Eumeces fasciatus*); it can grow to 8 inches (20 cm) in length. Widespread throughout the region, it favors moist woodlands. Another skink that you might encounter is the Broadhead Skink (*E. laticeps*). This skink also favors moist woodlands, and if you find its tracks you can probably also find the skink by sorting through the nearest pile of leaf litter or lifting up the nearest piece of wood. The Racerunner (*Cnemidophorus sexlineatus*), which can reach 11 inches (28 cm) in length, favors dry grasslands and well-drained woodlands. The smaller Fence Lizard (*Sceloporus undulatus*) is a possible find in sunny, open woodlands and grasslands.

Turtles

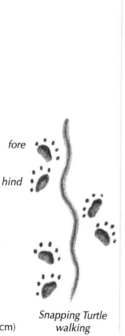

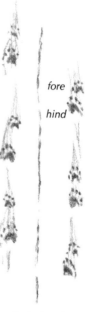

fore

hind

fore

hind

Straddle
4–10 in (10–25 cm)

*Snapping Turtle
walking*

*typical turtle
walking*

TURTLES

*Painted
Turtle*

Turtles, those ancient inhabitants of the water world, will happily slip into the murky depths to avoid detection. They do, however, come out from time to time to feed or to bask in the sunshine. Look for their distinctive tracks alongside ponds, rivers and moist areas. Some turtles, such as the huge Common Snapping Turtle (*Chelydra serpentina*), prefer to stay in the water and rarely come out.

With its large shell and short legs, a turtle leaves a track that has a straddle about half its body length. Although longer-legged turtles can raise their shells off the ground, short-legged species may let them drag, as shown in their tracks. The tail may leave a straight dragline in the mud. On firmer surfaces, look for distinct claw marks. One of the most widespread and prettiest turtles, often seen basking, is the Painted Turtle (*Chrysemys picta*); it can reach almost 10 inches (25 cm) in length. Slightly smaller is the well-named Common Musk Turtle (*Sternotherus odoratus*), which lives in shallow water and produces a foul odor. In southeastern Virginia, the Slider (*Trachemys scripta*), which reaches 12 inches (30 cm) in length, prefers slow-moving rivers. The Eastern Box Turtle (*Terrapene carolina*) is common in forested areas.

Snakes

SNAKES

Common Garter Snake

Many snake species inhabit this region, with much greater diversity in the warmer south. Because snakes are all long and slender, their tracks appear so similar that identification among the species is next to impossible. In fact, because a snake lacks feet and leaves a track that is just a gentle meander, it is very challenging even to establish in which direction a snake was moving.

The most frequently encountered snake, the harmless Common Garter Snake (*Thamnophis sirtalis*), can be found throughout the region, often near wet or moist areas; it can reach 4.3 feet (1.3 m) in length. Also widespread, but to only 16 inches (40 cm) in length, the Redbelly Snake (*Storeria occipitomaculata*) prefers hilly woodlands. The Rough Green Snake (*Opheodrys aestivus*), which can grow to 3.8 feet (1.2 m) long, inhabits grassy meadows and fields along forest edges. The rattlesnake most likely to be encountered is the Timber Rattlesnake (*Crotalus horridus*). It can grow to 6.3 feet (1.9 m) long, and it frequents a variety of habitats from marshlands to dry woodlands. Also common in a variety of habitats is the large and colorful Milk Snake (*Lampropeltis triangulum*), which grows up to 6.5 feet (2 m) long.

TRACK PATTERNS & PRINTS

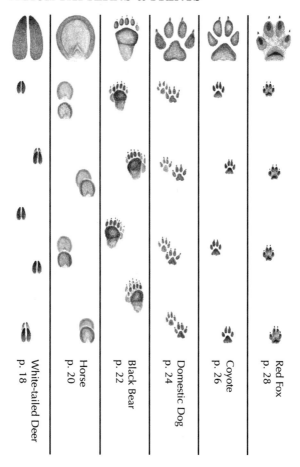

White-tailed Deer
p. 18

Horse
p. 20

Black Bear
p. 22

Domestic Dog
p. 24

Coyote
p. 26

Red Fox
p. 28

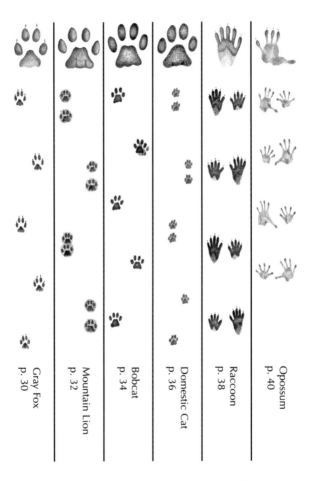

Gray Fox
p. 30

Mountain Lion
p. 32

Bobcat
p. 34

Domestic Cat
p. 36

Raccoon
p. 38

Opossum
p. 40

TRACK PATTERNS & PRINTS

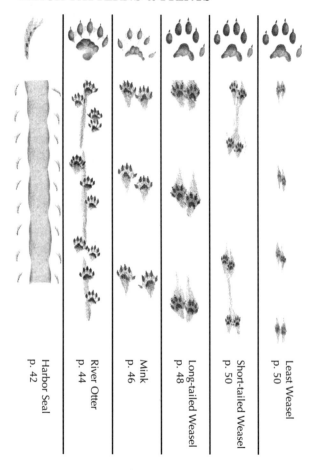

Harbor Seal
p. 42

River Otter
p. 44

Mink
p. 46

Long-tailed Weasel
p. 48

Short-tailed Weasel
p. 50

Least Weasel
p. 50

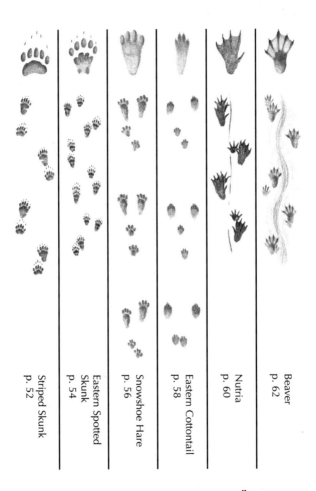

TRACK PATTERNS & PRINTS

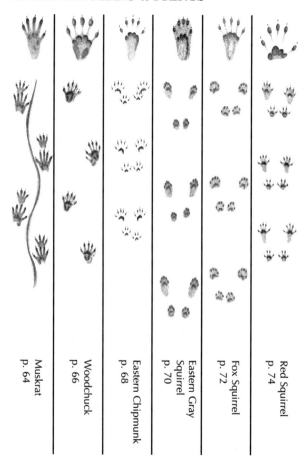

Muskrat
p. 64

Woodchuck
p. 66

Eastern Chipmunk
p. 68

Eastern Gray Squirrel
p. 70

Fox Squirrel
p. 72

Red Squirrel
p. 74

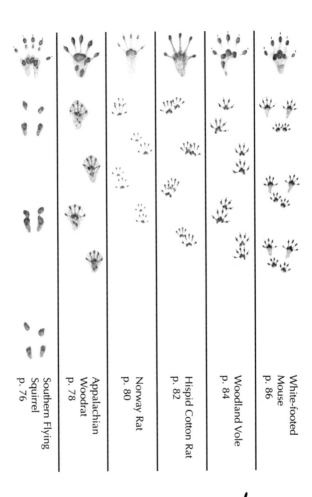

White-footed
Mouse
p. 86

Woodland Vole
p. 84

Hispid Cotton Rat
p. 82

Norway Rat
p. 80

Appalachian
Woodrat
p. 78

Southern Flying
Squirrel
p. 76

TRACK PATTERNS & PRINTS

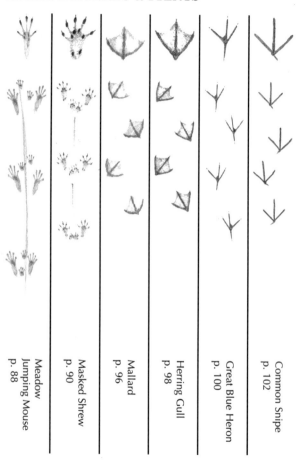

Meadow Jumping Mouse p. 88

Masked Shrew p. 90

Mallard p. 96

Herring Gull p. 98

Great Blue Heron p. 100

Common Snipe p. 102

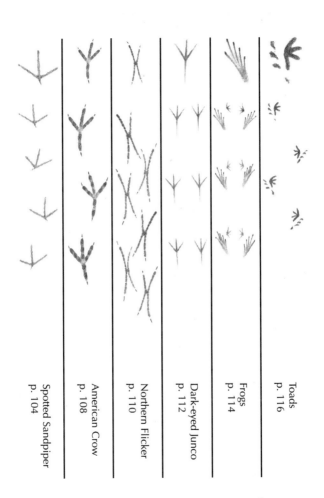

TRACK PATTERNS & PRINTS

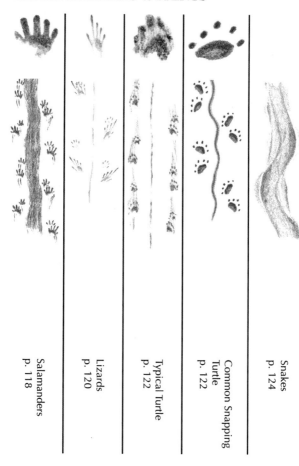

Salamanders
p. 118

Lizards
p. 120

Typical Turtle
p. 122

Common Snapping
Turtle
p. 122

Snakes
p. 124

HOOFED PRINTS

White-tailed
Deer

Horse

inch   cm

0 — 0

1

2 — 5

HIND PRINTS

White-footed
Mouse

Masked
Shrew

Hispid
Cotton Rat

Woodland
Vole

inch   cm

0 — 0

1

2

3

4

1

Southern
Flying Squirrel

Meadow
Jumping Mouse

Eastern
Chipmunk

2 — 5

135

HIND PRINTS

Norway
Rat

Appalachian
Woodrat

Red
Squirrel

Woodchuck

Muskrat

Fox
Squirrel

Eastern Gray
Squirrel

inch cm
0 ┬ 0

1

2 ┴ 5

Opossum

Raccoon

Eastern
Cottontail

HIND PRINTS

inch cm
0 ┬ 0

2

4 ┴ 10

Snowshoe
Hare

Nutria

Beaver

Black Bear

FORE PRINTS

Domestic
Cat

Bobcat

Gray
Fox

Red
Fox

Mountain Lion

Coyote

Domestic Dog

Least
Weasel

Short-tailed
Weasel

Long-tailed
Weasel

Mink

Eastern Spotted
Skunk

Striped
Skunk

River
Otter

inch   cm

0 — 0

1

2 — 5

BIBLIOGRAPHY

Behler, J.L., and F.W. King. 1979. *Field Guide to North American Reptiles and Amphibians.* National Audubon Society. New York: Alfred A. Knopf.

Brown, R., J. Ferguson, M. Lawrence and D. Lees. 1987. *Tracks and Signs of the Birds of Britain and Europe: An Identification Guide.* London: Christopher Helm.

Burt, W.H. 1976. *A Field Guide to the Mammals.* Boston: Houghton Mifflin Company.

Farrand, J., Jr. 1995. *Familiar Animal Tracks of North America.* National Audubon Society Pocket Guide. New York: Alfred A. Knopf.

Forrest, L.R. 1988. *Field Guide to Tracking Animals in Snow.* Harrisburg: Stackpole Books.

Halfpenny, J. 1986. *A Field Guide to Mammal Tracking in North America.* Boulder: Johnson Publishing Company.

Headstrom, R. 1971. *Identifying Animal Tracks.* Toronto: General Publishing Company.

Murie, O.J. 1974. *A Field Guide to Animal Tracks.* The Peterson Field Guide Series. Boston: Houghton Mifflin Company.

Rezendes, P. 1992. *Tracking and the Art of Seeing: How to Read Animal Tracks and Signs.* Vermont: Camden House Publishing.

Stall, C. 1989. *Animal Tracks of the Rocky Mountains.* Seattle: The Mountaineers.

Stokes, D., and L. Stokes. 1986. *A Guide to Animal Tracking and Behaviour.* Toronto: Little, Brown and Company.

Wassink, J.L. 1993. *Mammals of the Central Rockies.* Missoula: Mountain Press Publishing Company.

Whitaker, J.O., Jr. 1996. *National Audubon Society Field Guide to North American Mammals.* New York: Alfred A. Knopf.

INDEX

Page numbers in **boldface** type refer to the primary (illustrated) treatments of animal species and their tracks.

ABOUT THE AUTHOR

Tamara Eder, equipped from the age of six with a canoe, a dip net and a note pad, grew up with a fascination for nature and the diversity of life. She has a degree in environmental conservation sciences and has photographed and written about the biodiversity in Bermuda, the Galapagos Islands, the Amazon Basin, China, Tibet, Vietnam, Thailand and Malaysia.

Ian Sheldon, an accomplished artist, naturalist and educator, has lived in South Africa, Singapore, Britain and Canada. Caught collecting caterpillars at the age of three, he has been exposed to the beauty and diversity of nature ever since. He was educated at Cambridge University and the University of Alberta. When he is not in the tropics working on conservation projects or immersing himself in our beautiful wilderness, he is sharing his love for nature. Ian enjoys communicating this passion through the visual arts and the written word.